# THIS

*journal*

# BELONGS TO

_____

# DEDICATION

This book is dedicated to all the passionate and energetic beekeepers out there who are keeping the extreme sport of beekeeping alive!

You are my inspiration for producing books and I'm so excited to be a part of helping you plan for keeping your bees.

I hope you fulfill all your goals, hopes and dreams with this Beekeeping Journal!

Enjoy!

# HOW TO USE THIS BOOK

The purpose of this book is to keep all of your Beekeeping notes and information all in one place. It is a great way to record and keep track of your bees and will help keep you organized.

Here are some guidelines to follow so you can make the most out of using this book:

1. Today's Date - Write the date.
2. Colony Name - Record the name of the colony.
3. Hive Number - Write the hive number.
4. Queen Origin & Age - Where is your queen's origin and her age.
5. How Much Honey, Brood & Space? - Record how much of each.
6. Temperament - Check boxes to record whether the actions of the hive are Calm, Crazy or Pissy.
7. Population - Check boxes to record whether the hive is Low, Thriving or Normal.
8. Weather Conditions - Write what the weather conditions are.
9. Capacity - How full are the frames?
10. General Hive Appearance - Log your observations about the hive.
11. Signs Of Pests - Which pest is present, pest issues: Mites, Wax Moths, Ants, Dead Bees, Odor, Other
12. Is The Queen Present? - Do you see her actively entering or exiting the hive.
13. Is There One Egg Or Larva Per Cell? - Write your observation.
14. Notes Section- Enough blank lined space for writing any other important information.

# inspection notes

TODAY'S DATE:

COLONY NAME:

HIVE NUMBER:

QUEEN ORIGIN & AGE:

## How much

_____ Honey      _____ Brood      _____ Space

## Temperament

○ Calm          ○ Crazy          ○ Pissy

## Population

○ Low           ○ Thriving       ○ Normal

## Weather Conditions

_____

_____

## Capacity/How full are the frames?

_____

_____

## GENERAL HIVE APPEARANCE?

## Signs of Pests

- ◯ Mites
- ◯ Wax Moths
- ◯ Ants
- ◯ Dead Bees
- ◯ Odor
- ◯ Other

## Is the Queen present?

_____

## Is there one egg or larva per cell?

_____

## Notes

_____

_____

_____

_____

_____

# inspection notes

| |
|---|
| **TODAY'S DATE:** |
| **COLONY NAME:** |
| **HIVE NUMBER:** |
| **QUEEN ORIGIN & AGE:** |

## How much

_____ **Honey** _____ **Brood** _____ **Space**

## Temperament

◯ **Calm**          ◯ **Crazy**          ◯ **Pissy**

## Population

◯ **Low**          ◯ **Thriving**          ◯ **Normal**

## Weather Conditions

_____

_____

## Capacity/How full are the frames?

_____

_____

## GENERAL HIVE APPEARANCE?

## Signs of Pests

○ Mites          ○ Wax Moths          ○ Ants

○ Dead Bees     ○ Odor               ○ Other

## Is the Queen present?

_____

## Is there one egg or larva per cell?

_____

## Notes

_____

_____

_____

_____

_____

# inspection notes

TODAY'S DATE:
COLONY NAME:
HIVE NUMBER:
QUEEN ORIGIN & AGE:

## How much

_____ Honey _____ Brood _____ Space

## Temperament

○ Calm      ○ Crazy      ○ Pissy

## Population

○ Low      ○ Thriving      ○ Normal

## Weather Conditions

_____

_____

## Capacity/How full are the frames?

_____

_____

## GENERAL HIVE APPEARANCE?

## Signs of Pests

- ◯ Mites
- ◯ Wax Moths
- ◯ Ants
- ◯ Dead Bees
- ◯ Odor
- ◯ Other

## Is the Queen present?

_____

## Is there one egg or larva per cell?

_____

## Notes

_____

_____

_____

_____

_____

# inspection notes

TODAY'S DATE:

COLONY NAME:

HIVE NUMBER:

QUEEN ORIGIN & AGE:

## How much

_____ Honey      _____ Brood      _____ Space

## Temperament

○ Calm            ○ Crazy            ○ Pissy

## Population

○ Low             ○ Thriving         ○ Normal

## Weather Conditions

_____

_____

## Capacity/How full are the frames?

_____

_____

## GENERAL HIVE APPEARANCE?

### Signs of Pests

- ( ) Mites
- ( ) Dead Bees
- ( ) Wax Moths
- ( ) Odor
- ( ) Ants
- ( ) Other

### Is the Queen present?

_____

### Is there one egg or larva per cell?

_____

### Notes

_____

_____

_____

_____

_____

# inspection notes

TODAY'S DATE:

COLONY NAME:

HIVE NUMBER:

QUEEN ORIGIN & AGE:

## How much

_____ Honey   _____ Brood   _____ Space

## Temperament

○ Calm   ○ Crazy   ○ Pissy

## Population

○ Low   ○ Thriving   ○ Normal

## Weather Conditions

_____

_____

## Capacity/How full are the frames?

_____

_____

## GENERAL HIVE APPEARANCE?

## Signs of Pests

- ○ Mites
- ○ Wax Moths
- ○ Ants
- ○ Dead Bees
- ○ Odor
- ○ Other

## Is the Queen present?

_____

## Is there one egg or larva per cell?

_____

## Notes

_____

_____

_____

_____

_____

# inspection notes

**TODAY'S DATE:**
**COLONY NAME:**
**HIVE NUMBER:**
**QUEEN ORIGIN & AGE:**

## How much

_____ Honey  _____ Brood  _____ Space

## Temperament

◯ Calm        ◯ Crazy        ◯ Pissy

## Population

◯ Low         ◯ Thriving     ◯ Normal

## Weather Conditions

_____

_____

## Capacity/How full are the frames?

_____

_____

## GENERAL HIVE APPEARANCE?

## Signs of Pests

○ Mites      ○ Wax Moths      ○ Ants

○ Dead Bees      ○ Odor      ○ Other

## Is the Queen present?

_____

## Is there one egg or larva per cell?

_____

## Notes

_____

_____

_____

_____

_____

# inspection notes

| TODAY'S DATE: |
| COLONY NAME: |
| HIVE NUMBER: |
| QUEEN ORIGIN & AGE: |

## How much

_____ Honey  _____ Brood  _____ Space

## Temperament

○ Calm ○ Crazy ○ Pissy

## Population

○ Low ○ Thriving ○ Normal

## Weather Conditions

_____

_____

## Capacity/How full are the frames?

_____

_____

## GENERAL HIVE APPEARANCE?

## Signs of Pests

◯ Mites          ◯ Wax Moths          ◯ Ants

◯ Dead Bees      ◯ Odor               ◯ Other

## Is the Queen present?

_____

## Is there one egg or larva per cell?

_____

## Notes

_____

_____

_____

_____

_____

# inspection notes

| | |
|---|---|
| **TODAY'S DATE:** | |
| **COLONY NAME:** | |
| **HIVE NUMBER:** | |
| **QUEEN ORIGIN & AGE:** | |

## How much

_____ Honey    _____ Brood    _____ Space

## Temperament

○ Calm        ○ Crazy        ○ Pissy

## Population

○ Low         ○ Thriving     ○ Normal

## Weather Conditions

_____

_____

## Capacity/How full are the frames?

_____

_____

## GENERAL HIVE APPEARANCE?

## Signs of Pests

○ Mites  ○ Wax Moths  ○ Ants

○ Dead Bees  ○ Odor  ○ Other

## Is the Queen present?

## Is there one egg or larva per cell?

## Notes

# inspection notes

TODAY'S DATE:
COLONY NAME:
HIVE NUMBER:
QUEEN ORIGIN & AGE:

## How much

_____ Honey       _____ Brood       _____ Space

## Temperament

○ Calm              ○ Crazy              ○ Pissy

## Population

○ Low               ○ Thriving           ○ Normal

## Weather Conditions

_____

_____

## Capacity/How full are the frames?

_____

_____

## GENERAL HIVE APPEARANCE?

## Signs of Pests

○ Mites          ○ Wax Moths          ○ Ants

○ Dead Bees      ○ Odor               ○ Other

## Is the Queen present?

_____

## Is there one egg or larva per cell?

_____

## Notes

_____

_____

_____

_____

_____

# inspection notes

TODAY'S DATE:

COLONY NAME:

HIVE NUMBER:

QUEEN ORIGIN & AGE:

## How much

_____ Honey  _____ Brood  _____ Space

## Temperament

◯ Calm  ◯ Crazy  ◯ Pissy

## Population

◯ Low  ◯ Thriving  ◯ Normal

## Weather Conditions

_____

_____

## Capacity/How full are the frames?

_____

_____

## GENERAL HIVE APPEARANCE?

## Signs of Pests

○ Mites          ○ Wax Moths          ○ Ants

○ Dead Bees      ○ Odor               ○ Other

## Is the Queen present?

_____

## Is there one egg or larva per cell?

_____

## Notes

_____

_____

_____

_____

_____

_____

# inspection notes

TODAY'S DATE:

COLONY NAME:

HIVE NUMBER:

QUEEN ORIGIN & AGE:

## How much

_____ Honey _____ Brood _____ Space

## Temperament

○ Calm ○ Crazy ○ Pissy

## Population

○ Low ○ Thriving ○ Normal

## Weather Conditions

_____

_____

## Capacity/How full are the frames?

_____

_____

## GENERAL HIVE APPEARANCE?

## Signs of Pests

- ○ Mites
- ○ Dead Bees
- ○ Wax Moths
- ○ Odor
- ○ Ants
- ○ Other

## Is the Queen present?

_____

## Is there one egg or larva per cell?

_____

## Notes

_____

_____

_____

_____

_____

# inspection notes

TODAY'S DATE:
COLONY NAME:
HIVE NUMBER:
QUEEN ORIGIN & AGE:

## How much

_____ Honey _____ Brood _____ Space

## Temperament

○ Calm ○ Crazy ○ Pissy

## Population

○ Low ○ Thriving ○ Normal

## Weather Conditions

_____

_____

## Capacity/How full are the frames?

_____

_____

## GENERAL HIVE APPEARANCE?

### Signs of Pests

- ◯ Mites
- ◯ Dead Bees
- ◯ Wax Moths
- ◯ Odor
- ◯ Ants
- ◯ Other

### Is the Queen present?

_____

### Is there one egg or larva per cell?

_____

### Notes

_____

_____

_____

_____

_____

# inspection notes

TODAY'S DATE:

COLONY NAME:

HIVE NUMBER:

QUEEN ORIGIN & AGE:

## How much

_____ Honey          _____ Brood          _____ Space

## Temperament

○ Calm          ○ Crazy          ○ Pissy

## Population

○ Low          ○ Thriving          ○ Normal

## Weather Conditions

_____

_____

## Capacity/How full are the frames?

_____

_____

## GENERAL HIVE APPEARANCE?

### Signs of Pests

◯ Mites     ◯ Wax Moths     ◯ Ants

◯ Dead Bees     ◯ Odor     ◯ Other

### Is the Queen present?

_____

### Is there one egg or larva per cell?

_____

### Notes

_____

_____

_____

_____

_____

# inspection notes

TODAY'S DATE:

COLONY NAME:

HIVE NUMBER:

QUEEN ORIGIN & AGE:

## How much

_____ Honey  _____ Brood  _____ Space

## Temperament

○ Calm  ○ Crazy  ○ Pissy

## Population

○ Low  ○ Thriving  ○ Normal

## Weather Conditions

_____

_____

## Capacity/How full are the frames?

_____

_____

## GENERAL HIVE APPEARANCE?

### Signs of Pests

- ◯ Mites
- ◯ Dead Bees
- ◯ Wax Moths
- ◯ Odor
- ◯ Ants
- ◯ Other

### Is the Queen present?

_____

### Is there one egg or larva per cell?

_____

### Notes

_____

_____

_____

_____

_____

# inspection notes

TODAY'S DATE:
COLONY NAME:
HIVE NUMBER:
QUEEN ORIGIN & AGE:

## How much

_____ Honey    _____ Brood    _____ Space

## Temperament

◯ Calm            ◯ Crazy            ◯ Pissy

## Population

◯ Low             ◯ Thriving         ◯ Normal

## Weather Conditions

_____

_____

## Capacity/How full are the frames?

_____

_____

## GENERAL HIVE APPEARANCE?

### Signs of Pests

- ○ Mites
- ○ Dead Bees
- ○ Wax Moths
- ○ Odor
- ○ Ants
- ○ Other

### Is the Queen present?

_____

### Is there one egg or larva per cell?

_____

### Notes

_____

_____

_____

_____

_____

# inspection notes

**TODAY'S DATE:**

**COLONY NAME:**

**HIVE NUMBER:**

**QUEEN ORIGIN & AGE:**

## How much

_____ Honey _____ Brood _____ Space

## Temperament

○ Calm          ○ Crazy          ○ Pissy

## Population

○ Low          ○ Thriving          ○ Normal

## Weather Conditions

_____

_____

## Capacity/How full are the frames?

_____

_____

## GENERAL HIVE APPEARANCE?

## Signs of Pests

○ Mites          ○ Wax Moths          ○ Ants

○ Dead Bees      ○ Odor               ○ Other

## Is the Queen present?

_____

## Is there one egg or larva per cell?

_____

## Notes

_____

_____

_____

_____

_____

# inspection notes

**TODAY'S DATE:**

**COLONY NAME:**

**HIVE NUMBER:**

**QUEEN ORIGIN & AGE:**

## How much

_____ Honey _____ Brood _____ Space

## Temperament

○ Calm ○ Crazy ○ Pissy

## Population

○ Low ○ Thriving ○ Normal

## Weather Conditions

_____

_____

## Capacity/How full are the frames?

_____

_____

## GENERAL HIVE APPEARANCE?

## Signs of Pests

○ Mites          ○ Wax Moths          ○ Ants

○ Dead Bees      ○ Odor               ○ Other

## Is the Queen present?

_____

## Is there one egg or larva per cell?

_____

## Notes

_____

_____

_____

_____

_____

# inspection notes

**TODAY'S DATE:**

**COLONY NAME:**

**HIVE NUMBER:**

**QUEEN ORIGIN & AGE:**

## How much

_____ Honey _____ Brood _____ Space

## Temperament

○ Calm ○ Crazy ○ Pissy

## Population

○ Low ○ Thriving ○ Normal

## Weather Conditions

_____

_____

## Capacity/How full are the frames?

_____

_____

## GENERAL HIVE APPEARANCE?

## Signs of Pests

○ Mites      ○ Wax Moths      ○ Ants

○ Dead Bees      ○ Odor      ○ Other

## Is the Queen present?

_____

## Is there one egg or larva per cell?

_____

## Notes

_____

_____

_____

_____

_____

# inspection notes

TODAY'S DATE:
COLONY NAME:
HIVE NUMBER:
QUEEN ORIGIN & AGE:

## How much

_____ Honey    _____ Brood    _____ Space

## Temperament

○ Calm          ○ Crazy          ○ Pissy

## Population

○ Low           ○ Thriving       ○ Normal

## Weather Conditions

_____

_____

## Capacity/How full are the frames?

_____

_____

## GENERAL HIVE APPEARANCE?

## Signs of Pests

○ Mites      ○ Wax Moths      ○ Ants

○ Dead Bees      ○ Odor      ○ Other

## Is the Queen present?

_____

## Is there one egg or larva per cell?

_____

## Notes

_____

_____

_____

_____

_____

# inspection notes

**TODAY'S DATE:**

**COLONY NAME:**

**HIVE NUMBER:**

**QUEEN ORIGIN & AGE:**

## How much

_____ Honey _____ Brood _____ Space

## Temperament

◯ Calm          ◯ Crazy          ◯ Pissy

## Population

◯ Low          ◯ Thriving          ◯ Normal

## Weather Conditions

_____

_____

## Capacity/How full are the frames?

_____

_____

## GENERAL HIVE APPEARANCE?

## Signs of Pests

- ◯ Mites
- ◯ Wax Moths
- ◯ Ants
- ◯ Dead Bees
- ◯ Odor
- ◯ Other

## Is the Queen present?

_____

## Is there one egg or larva per cell?

_____

## Notes

_____

_____

_____

_____

_____

# inspection notes

**TODAY'S DATE:**

**COLONY NAME:**

**HIVE NUMBER:**

**QUEEN ORIGIN & AGE:**

## How much

_____ Honey  _____ Brood  _____ Space

## Temperament

◯ Calm  ◯ Crazy  ◯ Pissy

## Population

◯ Low  ◯ Thriving  ◯ Normal

## Weather Conditions

_____

_____

## Capacity/How full are the frames?

_____

_____

## GENERAL HIVE APPEARANCE?

## Signs of Pests

- ○ Mites
- ○ Dead Bees
- ○ Wax Moths
- ○ Odor
- ○ Ants
- ○ Other

## Is the Queen present?

_____

## Is there one egg or larva per cell?

_____

## Notes

_____

_____

_____

_____

_____

# inspection notes

**TODAY'S DATE:**

**COLONY NAME:**

**HIVE NUMBER:**

**QUEEN ORIGIN & AGE:**

## How much

_____ Honey _____ Brood _____ Space

## Temperament

○ Calm ○ Crazy ○ Pissy

## Population

○ Low ○ Thriving ○ Normal

## Weather Conditions

_____

_____

## Capacity/How full are the frames?

_____

_____

## GENERAL HIVE APPEARANCE?

## Signs of Pests

- ◯ Mites
- ◯ Wax Moths
- ◯ Ants
- ◯ Dead Bees
- ◯ Odor
- ◯ Other

## Is the Queen present?

_____

## Is there one egg or larva per cell?

_____

## Notes

_____

_____

_____

_____

_____

# inspection notes

TODAY'S DATE:
COLONY NAME:
HIVE NUMBER:
QUEEN ORIGIN & AGE:

## How much

_____ Honey _____ Brood _____ Space

## Temperament

◯ Calm          ◯ Crazy          ◯ Pissy

## Population

◯ Low          ◯ Thriving          ◯ Normal

## Weather Conditions

_____

_____

## Capacity/How full are the frames?

_____

_____

## GENERAL HIVE APPEARANCE?

## Signs of Pests

○ Mites     ○ Wax Moths     ○ Ants

○ Dead Bees     ○ Odor     ○ Other

## Is the Queen present?

_____

## Is there one egg or larva per cell?

_____

## Notes

_____

_____

_____

_____

_____

# inspection notes

TODAY'S DATE:

COLONY NAME:

HIVE NUMBER:

QUEEN ORIGIN & AGE:

## How much

_____ Honey     _____ Brood     _____ Space

## Temperament

○ Calm          ○ Crazy          ○ Pissy

## Population

○ Low           ○ Thriving       ○ Normal

## Weather Conditions

_____

_____

## Capacity/How full are the frames?

_____

_____

## GENERAL HIVE APPEARANCE?

### Signs of Pests

○ Mites     ○ Wax Moths     ○ Ants

○ Dead Bees     ○ Odor     ○ Other

### Is the Queen present?

_____

### Is there one egg or larva per cell?

_____

### Notes

_____

_____

_____

_____

_____

# inspection notes

**TODAY'S DATE:**
**COLONY NAME:**
**HIVE NUMBER:**
**QUEEN ORIGIN & AGE:**

## How much

_____ Honey        _____ Brood        _____ Space

## Temperament

◯ Calm              ◯ Crazy              ◯ Pissy

## Population

◯ Low               ◯ Thriving           ◯ Normal

## Weather Conditions

_____

_____

## Capacity/How full are the frames?

_____

_____

## GENERAL HIVE APPEARANCE?

### Signs of Pests

○ Mites      ○ Wax Moths      ○ Ants

○ Dead Bees      ○ Odor      ○ Other

### Is the Queen present?

_____

### Is there one egg or larva per cell?

_____

### Notes

_____

_____

_____

_____

_____

# inspection notes

TODAY'S DATE:

COLONY NAME:

HIVE NUMBER:

QUEEN ORIGIN & AGE:

## How much

_____ Honey    _____ Brood    _____ Space

## Temperament

○ Calm          ○ Crazy          ○ Pissy

## Population

○ Low           ○ Thriving       ○ Normal

## Weather Conditions

_____

_____

## Capacity/How full are the frames?

_____

_____

## GENERAL HIVE APPEARANCE?

## Signs of Pests

○ Mites      ○ Wax Moths      ○ Ants

○ Dead Bees      ○ Odor      ○ Other

## Is the Queen present?

_____

## Is there one egg or larva per cell?

_____

## Notes

_____

_____

_____

_____

_____

# inspection notes

TODAY'S DATE:

COLONY NAME:

HIVE NUMBER:

QUEEN ORIGIN & AGE:

## How much

_____ Honey     _____ Brood     _____ Space

## Temperament

○ Calm     ○ Crazy     ○ Pissy

## Population

○ Low     ○ Thriving     ○ Normal

## Weather Conditions

_____

_____

## Capacity/How full are the frames?

_____

_____

## GENERAL HIVE APPEARANCE?

### Signs of Pests

○ Mites　　○ Wax Moths　　○ Ants

○ Dead Bees　　○ Odor　　○ Other

### Is the Queen present?

_____

### Is there one egg or larva per cell?

_____

### Notes

_____

_____

_____

_____

_____

# inspection notes

**TODAY'S DATE:**

**COLONY NAME:**

**HIVE NUMBER:**

**QUEEN ORIGIN & AGE:**

## How much

_____ Honey  _____ Brood  _____ Space

## Temperament

○ Calm  ○ Crazy  ○ Pissy

## Population

○ Low  ○ Thriving  ○ Normal

## Weather Conditions

_____

_____

## Capacity/How full are the frames?

_____

_____

## GENERAL HIVE APPEARANCE?

## Signs of Pests

- ○ Mites
- ○ Dead Bees
- ○ Wax Moths
- ○ Odor
- ○ Ants
- ○ Other

## Is the Queen present?

_____

## Is there one egg or larva per cell?

_____

## Notes

_____

_____

_____

_____

_____

# inspection notes

TODAY'S DATE:

COLONY NAME:

HIVE NUMBER:

QUEEN ORIGIN & AGE:

## How much

_____ Honey    _____ Brood    _____ Space

## Temperament

○ Calm    ○ Crazy    ○ Pissy

## Population

○ Low    ○ Thriving    ○ Normal

## Weather Conditions

_____

_____

## Capacity/How full are the frames?

_____

_____

## GENERAL HIVE APPEARANCE?

### Signs of Pests

- ◯ Mites
- ◯ Dead Bees
- ◯ Wax Moths
- ◯ Odor
- ◯ Ants
- ◯ Other

### Is the Queen present?

_____

### Is there one egg or larva per cell?

_____

### Notes

_____

_____

_____

_____

_____

# inspection notes

**TODAY'S DATE:**

**COLONY NAME:**

**HIVE NUMBER:**

**QUEEN ORIGIN & AGE:**

## How much

_____ Honey _____ Brood _____ Space

## Temperament

○ Calm ○ Crazy ○ Pissy

## Population

○ Low ○ Thriving ○ Normal

## Weather Conditions

_____

_____

## Capacity/How full are the frames?

_____

_____

## GENERAL HIVE APPEARANCE?

## Signs of Pests

- ◯ Mites
- ◯ Wax Moths
- ◯ Ants
- ◯ Dead Bees
- ◯ Odor
- ◯ Other

## Is the Queen present?

_____

## Is there one egg or larva per cell?

_____

## Notes

_____

_____

_____

_____

_____

# inspection notes

**TODAY'S DATE:**

**COLONY NAME:**

**HIVE NUMBER:**

**QUEEN ORIGIN & AGE:**

## How much

_____ Honey    _____ Brood    _____ Space

## Temperament

○ Calm          ○ Crazy          ○ Pissy

## Population

○ Low           ○ Thriving       ○ Normal

## Weather Conditions

_____

_____

## Capacity/How full are the frames?

_____

_____

## GENERAL HIVE APPEARANCE?

## Signs of Pests

- ⊙ Mites
- ⊙ Wax Moths
- ⊙ Ants
- ⊙ Dead Bees
- ⊙ Odor
- ⊙ Other

## Is the Queen present?

## Is there one egg or larva per cell?

## Notes

# inspection notes

TODAY'S DATE:

COLONY NAME:

HIVE NUMBER:

QUEEN ORIGIN & AGE:

## How much

_____ Honey _____ Brood _____ Space

## Temperament

◯ Calm ◯ Crazy ◯ Pissy

## Population

◯ Low ◯ Thriving ◯ Normal

## Weather Conditions

_____

_____

## Capacity/How full are the frames?

_____

_____

## GENERAL HIVE APPEARANCE?

## Signs of Pests

- ◯ Mites
- ◯ Dead Bees
- ◯ Wax Moths
- ◯ Odor
- ◯ Ants
- ◯ Other

## Is the Queen present?

_____

## Is there one egg or larva per cell?

_____

## Notes

_____

_____

_____

_____

_____

# inspection notes

| |
|---|
| **TODAY'S DATE:** |
| **COLONY NAME:** |
| **HIVE NUMBER:** |
| **QUEEN ORIGIN & AGE:** |

## How much

_____ Honey       _____ Brood       _____ Space

## Temperament

○ Calm          ○ Crazy          ○ Pissy

## Population

○ Low          ○ Thriving          ○ Normal

## Weather Conditions

_____

_____

## Capacity/How full are the frames?

_____

_____

## GENERAL HIVE APPEARANCE?

## Signs of Pests

- ( ) Mites
- ( ) Dead Bees
- ( ) Wax Moths
- ( ) Odor
- ( ) Ants
- ( ) Other

## Is the Queen present?

_____

## Is there one egg or larva per cell?

_____

## Notes

_____

_____

_____

_____

_____

# inspection notes

TODAY'S DATE:

COLONY NAME:

HIVE NUMBER:

QUEEN ORIGIN & AGE:

## How much

_____ Honey _____ Brood _____ Space

## Temperament

○ Calm          ○ Crazy          ○ Pissy

## Population

○ Low          ○ Thriving          ○ Normal

## Weather Conditions

_____

_____

## Capacity/How full are the frames?

_____

_____

## GENERAL HIVE APPEARANCE?

## Signs of Pests

- ◯ Mites
- ◯ Wax Moths
- ◯ Ants
- ◯ Dead Bees
- ◯ Odor
- ◯ Other

## Is the Queen present?

_____

## Is there one egg or larva per cell?

_____

## Notes

_____

_____

_____

_____

_____

# inspection notes

| TODAY'S DATE: |
| COLONY NAME: |
| HIVE NUMBER: |
| QUEEN ORIGIN & AGE: |

## How much

_____ Honey    _____ Brood    _____ Space

## Temperament

○ Calm          ○ Crazy          ○ Pissy

## Population

○ Low           ○ Thriving       ○ Normal

## Weather Conditions

_____

_____

## Capacity/How full are the frames?

_____

_____

## GENERAL HIVE APPEARANCE?

### Signs of Pests

○ Mites          ○ Wax Moths          ○ Ants

○ Dead Bees      ○ Odor               ○ Other

### Is the Queen present?

_____

### Is there one egg or larva per cell?

_____

### Notes

_____

_____

_____

_____

_____

# inspection notes

**TODAY'S DATE:**

**COLONY NAME:**

**HIVE NUMBER:**

**QUEEN ORIGIN & AGE:**

## How much

_____ Honey _____ Brood _____ Space

## Temperament

○ Calm ○ Crazy ○ Pissy

## Population

○ Low ○ Thriving ○ Normal

## Weather Conditions

_____

_____

## Capacity/How full are the frames?

_____

_____

## GENERAL HIVE APPEARANCE?

## Signs of Pests

◯ Mites  ◯ Wax Moths  ◯ Ants

◯ Dead Bees  ◯ Odor  ◯ Other

## Is the Queen present?

_____

## Is there one egg or larva per cell?

_____

## Notes

_____

_____

_____

_____

_____

# inspection notes

TODAY'S DATE:

COLONY NAME:

HIVE NUMBER:

QUEEN ORIGIN & AGE:

## How much

_____ Honey _____ Brood _____ Space

## Temperament

○ Calm ○ Crazy ○ Pissy

## Population

○ Low ○ Thriving ○ Normal

## Weather Conditions

_____

_____

## Capacity/How full are the frames?

_____

_____

## GENERAL HIVE APPEARANCE?

## Signs of Pests

- ◯ Mites
- ◯ Wax Moths
- ◯ Ants
- ◯ Dead Bees
- ◯ Odor
- ◯ Other

## Is the Queen present?

_____

## Is there one egg or larva per cell?

_____

## Notes

_____

_____

_____

_____

_____

# inspection notes

TODAY'S DATE:

COLONY NAME:

HIVE NUMBER:

QUEEN ORIGIN & AGE:

## How much

_____ Honey _____ Brood _____ Space

## Temperament

○ Calm ○ Crazy ○ Pissy

## Population

○ Low ○ Thriving ○ Normal

## Weather Conditions

_____

_____

## Capacity/How full are the frames?

_____

_____

## GENERAL HIVE APPEARANCE?

## Signs of Pests

- ◯ Mites
- ◯ Dead Bees
- ◯ Wax Moths
- ◯ Odor
- ◯ Ants
- ◯ Other

## Is the Queen present?

___

## Is there one egg or larva per cell?

___

## Notes

___

___

___

___

# inspection notes

TODAY'S DATE:

COLONY NAME:

HIVE NUMBER:

QUEEN ORIGIN & AGE:

## How much

_____ Honey _____ Brood _____ Space

## Temperament

○ Calm ○ Crazy ○ Pissy

## Population

○ Low ○ Thriving ○ Normal

## Weather Conditions

_____

_____

## Capacity/How full are the frames?

_____

_____

## GENERAL HIVE APPEARANCE?

### Signs of Pests

- ◯ Mites
- ◯ Dead Bees
- ◯ Wax Moths
- ◯ Odor
- ◯ Ants
- ◯ Other

### Is the Queen present?

_____

### Is there one egg or larva per cell?

_____

### Notes

_____

_____

_____

_____

_____

_____

# inspection notes

TODAY'S DATE:

COLONY NAME:

HIVE NUMBER:

QUEEN ORIGIN & AGE:

## How much

_____ Honey   _____ Brood   _____ Space

## Temperament

○ Calm          ○ Crazy          ○ Pissy

## Population

○ Low           ○ Thriving       ○ Normal

## Weather Conditions

_____

_____

## Capacity/How full are the frames?

_____

_____

## GENERAL HIVE APPEARANCE?

### Signs of Pests

- ○ Mites
- ○ Dead Bees
- ○ Wax Moths
- ○ Odor
- ○ Ants
- ○ Other

### Is the Queen present?

_____

### Is there one egg or larva per cell?

_____

### Notes

_____

_____

_____

_____

_____

# inspection notes

**TODAY'S DATE:**

**COLONY NAME:**

**HIVE NUMBER:**

**QUEEN ORIGIN & AGE:**

## How much

_____ Honey     _____ Brood     _____ Space

## Temperament

○ Calm     ○ Crazy     ○ Pissy

## Population

○ Low     ○ Thriving     ○ Normal

## Weather Conditions

_____

_____

## Capacity/How full are the frames?

_____

_____

## GENERAL HIVE APPEARANCE?

### Signs of Pests

○ Mites     ○ Wax Moths     ○ Ants

○ Dead Bees     ○ Odor     ○ Other

### Is the Queen present?

_____

### Is there one egg or larva per cell?

_____

### Notes

_____

_____

_____

_____

_____

# inspection notes

TODAY'S DATE:

COLONY NAME:

HIVE NUMBER:

QUEEN ORIGIN & AGE:

## How much

_____ Honey  _____ Brood  _____ Space

## Temperament

○ Calm  ○ Crazy  ○ Pissy

## Population

○ Low  ○ Thriving  ○ Normal

## Weather Conditions

_____

_____

## Capacity/How full are the frames?

_____

_____

## GENERAL HIVE APPEARANCE?

## Signs of Pests

○ Mites          ○ Wax Moths          ○ Ants

○ Dead Bees      ○ Odor               ○ Other

## Is the Queen present!

---

## Is there one egg or larva per cell?

---

## Notes

---

---

---

---

---

# inspection notes

TODAY'S DATE:

COLONY NAME:

HIVE NUMBER:

QUEEN ORIGIN & AGE:

## How much

_____ Honey _____ Brood _____ Space

## Temperament

◯ Calm ◯ Crazy ◯ Pissy

## Population

◯ Low ◯ Thriving ◯ Normal

## Weather Conditions

_____

_____

## Capacity/How full are the frames?

_____

_____

## GENERAL HIVE APPEARANCE?

## Signs of Pests

○ Mites     ○ Wax Moths     ○ Ants

○ Dead Bees     ○ Odor     ○ Other

## Is the Queen present?

_____

## Is there one egg or larva per cell?

_____

## Notes

_____

_____

_____

_____

_____

# inspection notes

**TODAY'S DATE:**

**COLONY NAME:**

**HIVE NUMBER:**

**QUEEN ORIGIN & AGE:**

## How much

_____ **Honey**     _____ **Brood**     _____ **Space**

## Temperament

◯ **Calm**          ◯ **Crazy**          ◯ **Pissy**

## Population

◯ **Low**          ◯ **Thriving**          ◯ **Normal**

## Weather Conditions

_____

_____

## Capacity/How full are the frames?

_____

_____

## GENERAL HIVE APPEARANCE?

## Signs of Pests

- ○ Mites
- ○ Dead Bees
- ○ Wax Moths
- ○ Odor
- ○ Ants
- ○ Other

## Is the Queen present?

_____

## Is there one egg or larva per cell?

_____

## Notes

_____

_____

_____

_____

_____

# inspection notes

| |
|---|
| **TODAY'S DATE:** |
| **COLONY NAME:** |
| **HIVE NUMBER:** |
| **QUEEN ORIGIN & AGE:** |

## How much

_____ Honey        _____ Brood        _____ Space

## Temperament

○ Calm        ○ Crazy        ○ Pissy

## Population

○ Low        ○ Thriving        ○ Normal

## Weather Conditions

_____

_____

## Capacity/How full are the frames?

_____

_____

## GENERAL HIVE APPEARANCE?

## Signs of Pests

○ Mites     ○ Wax Moths     ○ Ants

○ Dead Bees     ○ Odor     ○ Other

## Is the Queen present?

_____

## Is there one egg or larva per cell?

_____

## Notes

_____

_____

_____

_____

_____

# inspection notes

TODAY'S DATE:

COLONY NAME:

HIVE NUMBER:

QUEEN ORIGIN & AGE:

## How much

_____ Honey     _____ Brood     _____ Space

## Temperament

○ Calm     ○ Crazy     ○ Pissy

## Population

○ Low     ○ Thriving     ○ Normal

## Weather Conditions

_____

_____

## Capacity/How full are the frames?

_____

_____

## GENERAL HIVE APPEARANCE?

## Signs of Pests

- ○ Mites
- ○ Dead Bees
- ○ Wax Moths
- ○ Odor
- ○ Ants
- ○ Other

## Is the Queen present?

_____

## Is there one egg or larva per cell?

_____

## Notes

_____

_____

_____

_____

_____

# inspection notes

TODAY'S DATE:

COLONY NAME:

HIVE NUMBER:

QUEEN ORIGIN & AGE:

## How much

_____ Honey    _____ Brood    _____ Space

## Temperament

◯ Calm          ◯ Crazy          ◯ Pissy

## Population

◯ Low           ◯ Thriving       ◯ Normal

## Weather Conditions

_____

_____

## Capacity/How full are the frames?

_____

_____

## GENERAL HIVE APPEARANCE?

## Signs of Pests

- ○ Mites
- ○ Wax Moths
- ○ Ants
- ○ Dead Bees
- ○ Odor
- ○ Other

## Is the Queen present?

_____

## Is there one egg or larva per cell?

_____

## Notes

_____

_____

_____

_____

_____

# inspection notes

TODAY'S DATE:

COLONY NAME:

HIVE NUMBER:

QUEEN ORIGIN & AGE:

## How much

_____ Honey _____ Brood _____ Space

## Temperament

○ Calm ○ Crazy ○ Pissy

## Population

○ Low ○ Thriving ○ Normal

## Weather Conditions

_____

_____

## Capacity/How full are the frames?

_____

_____

## GENERAL HIVE APPEARANCE?

## Signs of Pests

- ○ Mites
- ○ Wax Moths
- ○ Ants
- ○ Dead Bees
- ○ Odor
- ○ Other

## Is the Queen present?

_____

## Is there one egg or larva per cell?

_____

## Notes

_____

_____

_____

_____

# inspection notes

TODAY'S DATE:

COLONY NAME:

HIVE NUMBER:

QUEEN ORIGIN & AGE:

## How much

_____ Honey    _____ Brood    _____ Space

## Temperament

○ Calm    ○ Crazy    ○ Pissy

## Population

○ Low    ○ Thriving    ○ Normal

## Weather Conditions

_____

_____

## Capacity/How full are the frames?

_____

_____

## GENERAL HIVE APPEARANCE?

## Signs of Pests

○ Mites     ○ Wax Moths     ○ Ants

○ Dead Bees     ○ Odor     ○ Other

## Is the Queen present?

_____

## Is there one egg or larva per cell?

_____

## Notes

_____

_____

_____

_____

_____

# inspection notes

**TODAY'S DATE:**
**COLONY NAME:**
**HIVE NUMBER:**
**QUEEN ORIGIN & AGE:**

## How much

_____ Honey _____ Brood _____ Space

## Temperament

○ Calm ○ Crazy ○ Pissy

## Population

○ Low ○ Thriving ○ Normal

## Weather Conditions

_____

_____

## Capacity/How full are the frames?

_____

_____

## GENERAL HIVE APPEARANCE?

## Signs of Pests

- ◯ Mites
- ◯ Dead Bees
- ◯ Wax Moths
- ◯ Odor
- ◯ Ants
- ◯ Other

## Is the Queen present?

_____

## Is there one egg or larva per cell?

_____

## Notes

_____

_____

_____

_____

_____

# inspection notes

TODAY'S DATE:

COLONY NAME:

HIVE NUMBER:

QUEEN ORIGIN & AGE:

## How much

_____ Honey   _____ Brood   _____ Space

## Temperament

○ Calm          ○ Crazy          ○ Pissy

## Population

○ Low           ○ Thriving       ○ Normal

## Weather Conditions

_____

_____

## Capacity/How full are the frames?

_____

_____

## GENERAL HIVE APPEARANCE?

## Signs of Pests

- ○ Mites
- ○ Dead Bees
- ○ Wax Moths
- ○ Odor
- ○ Ants
- ○ Other

## Is the Queen present?

_____

## Is there one egg or larva per cell?

_____

## Notes

_____

_____

_____

_____

_____

# inspection notes

TODAY'S DATE:

COLONY NAME:

HIVE NUMBER:

QUEEN ORIGIN & AGE:

## How much

_____ Honey _____ Brood _____ Space

## Temperament

⚪ Calm ⚪ Crazy ⚪ Pissy

## Population

⚪ Low ⚪ Thriving ⚪ Normal

## Weather Conditions

_____

_____

## Capacity/How full are the frames?

_____

_____

## GENERAL HIVE APPEARANCE?

## Signs of Pests

○ Mites  ○ Wax Moths  ○ Ants

○ Dead Bees  ○ Odor  ○ Other

## Is the Queen present?

_____

## Is there one egg or larva per cell?

_____

## Notes

_____

_____

_____

_____

_____

# inspection notes

TODAY'S DATE:

COLONY NAME:

HIVE NUMBER:

QUEEN ORIGIN & AGE:

## How much

_____ Honey _____ Brood _____ Space

## Temperament

○ Calm ○ Crazy ○ Pissy

## Population

○ Low ○ Thriving ○ Normal

## Weather Conditions

_____

_____

## Capacity/How full are the frames?

_____

_____

## GENERAL HIVE APPEARANCE?

## Signs of Pests

◯ Mites       ◯ Wax Moths       ◯ Ants

◯ Dead Bees   ◯ Odor            ◯ Other

## Is the Queen present?

_____

## Is there one egg or larva per cell?

_____

## Notes

_____

_____

_____

_____

_____

# inspection notes

TODAY'S DATE:

COLONY NAME:

HIVE NUMBER:

QUEEN ORIGIN & AGE:

## How much

_____ Honey        _____ Brood        _____ Space

## Temperament

○ Calm            ○ Crazy            ○ Pissy

## Population

○ Low             ○ Thriving         ○ Normal

## Weather Conditions

_____

_____

## Capacity/How full are the frames?

_____

_____

CPSIA information can be obtained
at www.ICGtesting.com
Printed in the USA
LVHW101110270520
656347LV00014B/96